Hot, Hot, Hot

written by Maryann Dobeck
illustrated by Shelley Dieterichs

THIS BOOK IS THE PROPERTY OF:

STATE ____ PROVINCE ____ COUNTY ____ PARISH ____ SCHOOL DISTRICT ____ OTHER ____	**Book No.** ____ **Enter information in spaces to the left as instructed**

ISSUED TO	*Year Used*	*CONDITION*	
		ISSUED	*RETURNED*

PUPILS to whom this textbook is issued must not write on any page or mark any part of it in any way, consumable textbooks excepted.

1. Teachers should see that the pupil's name is clearly written in ink in the spaces above in every book issued.
2. The following terms should be used in recording the condition of the book: New; Good; Fair; Poor; Bad.

Pal got hot.

Hot, hot, hot!

A log got hot.

Hot, hot, hot!

A pot got hot.

Hot, hot, hot!

Pop! Pop! Pop!

The End

Understanding the Story

Questions are to be read aloud by a teacher or parent.

1. What is the title of this story?
2. What got hot in the story?
3. What popped out of the pot?

Answers: 1. Hot, Hot, Hot 2. a dog, a log, and a pot 3. popcorn

---

Saxon Publishers, Inc.
Editorial: Barbara Place, Julie Webster, Grey Allman, Elisha Mayer
Production: Angela Johnson, Carrie Brown, Cristi Henderson

Brown Publishing Network, Inc.
Editorial: Marie Brown, Gale Clifford, Maryann Dobeck
Art/Design: Trelawney Goodell, Camille Venti, Joan Asikainen
Production: Joseph Hinckley

Published by Harcourt Achieve Inc.

Saxon is a trademark of Harcourt Achieve Inc.

Printed in the United States of America
ISBN: 1-56577-947-9

6 7 8 546 12 11 10 09 08

Phonetic Concepts Practiced

l (Pal)
ŏ (log)
g (got)
h (hot)
t (pot)
p (Pop!)
ă (at)
the word "a" (A log)

Nondecodable Sight Words Introduced

The End

ISBN 1-56577-947-9

Grade K, Decodable Reade
First used in Lesson 27